Gorilla, Baboon, and Other Troops

Heather Hudak

Crabtree Publishing
crabtreebooks.com

Author: Heather Hudak

Series research and development:
Reagan Miller, Janine Deschenes

Editor: Janine Deschenes

Proofreader: Kathy Middleton

Design and photo research:
Katherine Berti

Print and production coordinator:
Katherine Berti

Image credits:
AgeFotostock
Anup Shah: p. 22 (bottom)
Alamy Stock Photo
David South: p. 20–21 (bottom)
All other images from Shutterstock

Crabtree Publishing

crabtreebooks.com 800-387-7650

 In Canada: We acknowledge the financial support of the Government of Canada through the Canada Book Fund for our publishing activities.

Hardcover 978-1-0398-0653-5
Paperback 978-1-0398-0679-5
Ebook (pdf) 978-1-0398-0704-4
Epub 978-1-0398-0731-0

Published in Canada
Crabtree Publishing
616 Welland Avenue
St. Catharines, Ontario
L2M 5V6

Published in the United States
Crabtree Publishing
347 Fifth Avenue
Suite 1402-145
New York, New York, 10016

Library and Archives Canada
Cataloguing in Publication
Available at Library and Archives Canada

Library of Congress
Cataloging-in-Publication Data
Available at the Library of Congress

Printed in the U.S.A./012023/CG20220815

Contents

What is a Troop?

What would life be like if you lived alone, without your family or friends? It would be a lot harder to get the things you need to survive. It might also be a lot harder to have fun! Humans live and interact in groups with their family members and friends. So do many animals!

Each member of a troop has a job to do. Some members raise the young. Others search for food or protect the rest of the troop from ***predators****.*

A troop stays within its ***home range****. It does not often leave or meet up with other troops.*

There are different types of animal groups. Some are called troops. A troop is a community of animals that work together for survival. Troops can have a few members or a few thousand members! All animals in the troop belong to the same animal **species**. Most **primates**, such as gorillas and baboons, live in troops.

Some other animal species, such as ostriches, also live in troops.

Why Do Animals Live in Troops?

Life in the wild can be hard. Being part of a troop makes it easier to survive. Animals in a troop can work together to find food and stay safe.

Many animals in a troop can look for food at the same time. They let each other know when they find something to eat. Troop members also work together to find water and places to sleep. Staying safe from predators is another reason animals form troops. Animals that live alone are more likely to become **prey**. Predators cannot attack an entire troop at the same time.

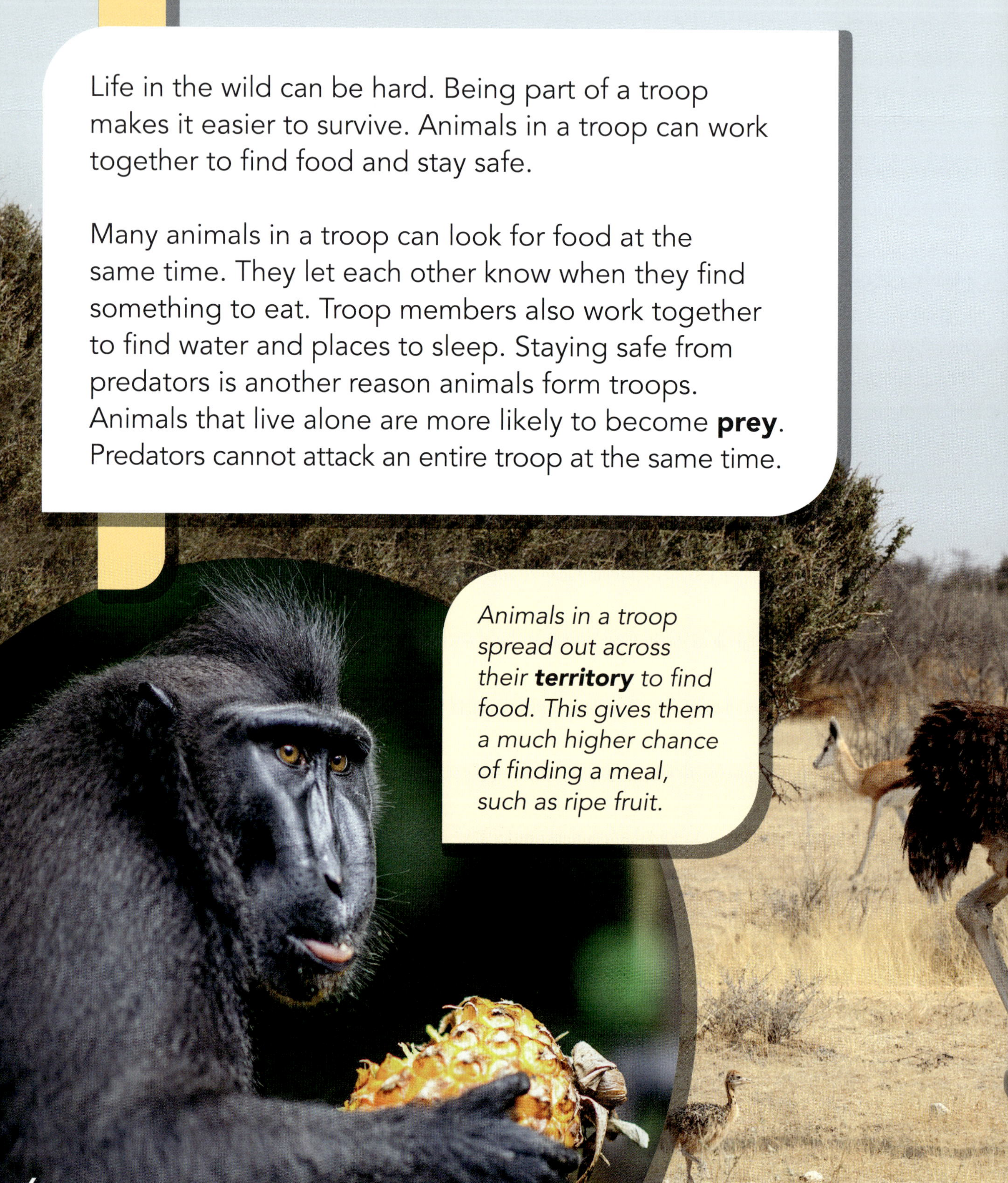

*Animals in a troop spread out across their **territory** to find food. This gives them a much higher chance of finding a meal, such as ripe fruit.*

Sticking Together

For animals that live alone, it is not always easy to find an animal to **mate** with. But animals that live in troops can mate with other animals in their troop.

Members of the troop work together to ***deter*** *predators and defend their home range.*

Gorillas

At a Glance

Number of species: Two—eastern and western
Size: Two to more than 50 members
Location: Tropical forests of Africa

Gorillas are the largest primates in the world. They have thick, strong limbs and broad chests with wide shoulders. Males weigh up to around 485 pounds (220 kg)! Despite their size, gorillas are some of the most peaceful, gentle animals on the planet.

Gorillas live in family groups. These troops are made up of one or two adult males, a few younger males, and many females and their offspring. The oldest, strongest male leads the group. He is responsible for the troop's health and safety. He makes all decisions for the group, such as when to eat and sleep.

Adult male gorillas are called silverbacks. This name comes from the gray or silver hairs on their lower backs.

Did You Know?

Gorillas are closely related to humans and often behave like them. They are very smart and can show different emotions, such as happiness and sadness.

Gorillas are up to 10 times stronger than most humans. They can easily tear down banana trees!

Life in a Troop

*Gorillas play important roles in their **ecosystems**. They spread the seeds of the plants they eat in their **scat**. They also create spaces between trees when they move around. This lets the Sun shine through and reach smaller plants below.*

Most gorilla troops have five to 10 members. Females leave their troops when they are about 10 years old. They join new troops and mate with the **dominant** males. They might change troops many times.

Most silverbacks leave their troops when they are about nine years old. They spend a few years on their own. Then, they form new troops or try to replace a silverback in another troop. Sometimes, a silverback might stay with his birth troop for his entire life. He might lead alongside the dominant male if it is his father or brother.

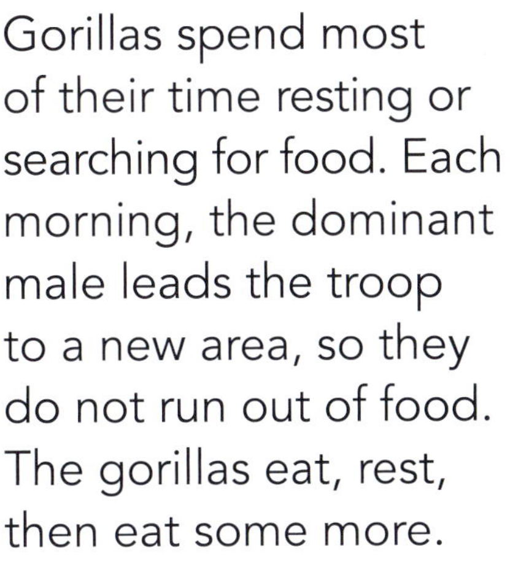

Gorillas spend most of their time resting or searching for food. Each morning, the dominant male leads the troop to a new area, so they do not run out of food. The gorillas eat, rest, then eat some more.

A gorilla might scare away enemies by breaking off tree branches and shaking them.

Sticking Together

Gorillas might bark, roar, or scream. But they are mostly quiet animals. They often use body language to express themselves. They crouch down when they want to show they mean no harm. They walk tall when they feel **confident**. They slap their chests when they are angry.

Gorillas mainly eat leafy plants, bamboo shoots, and fruit. They eat as much as 66 pounds (30 kg) of food each day.

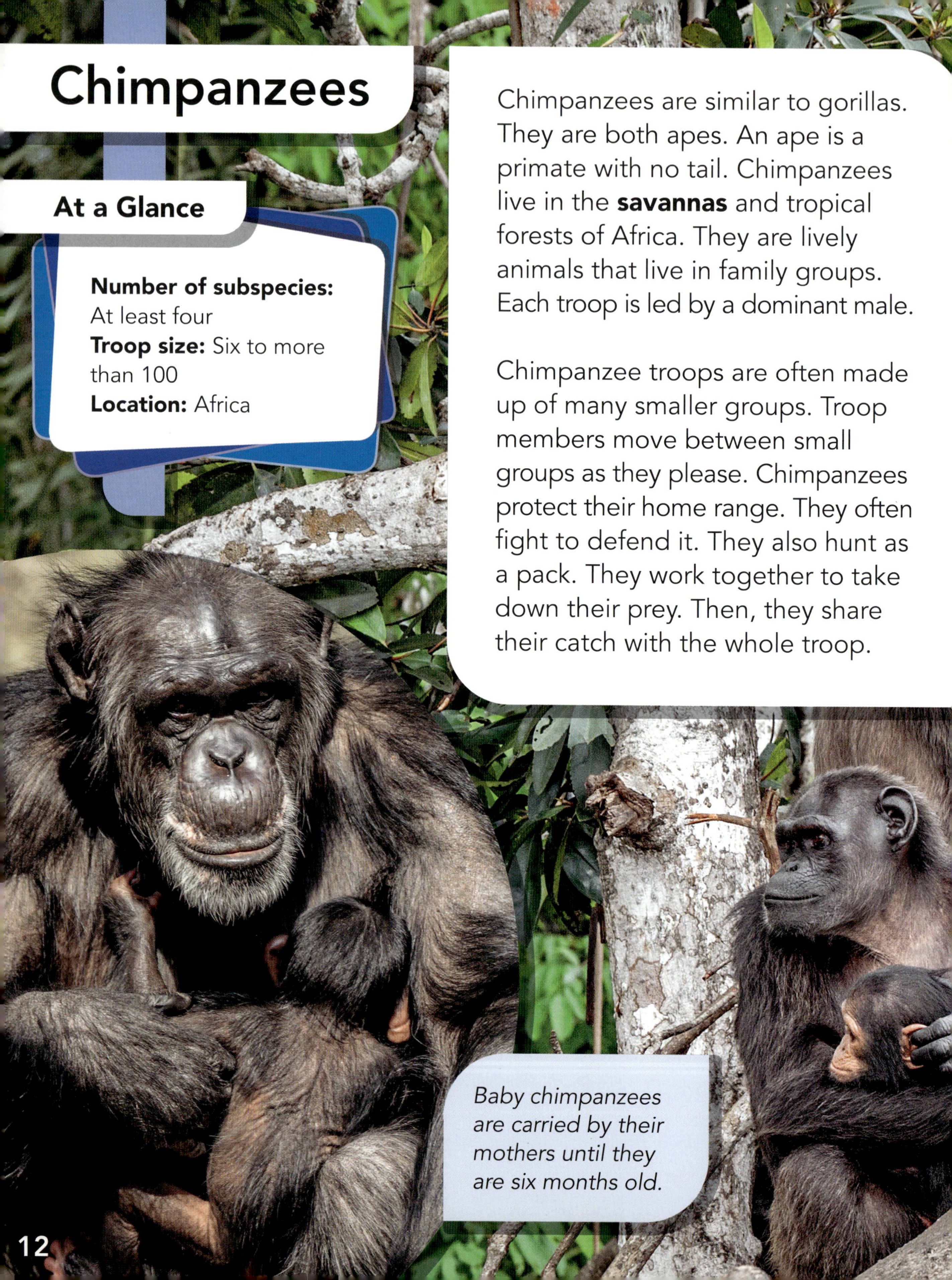

Chimpanzees

At a Glance

Number of subspecies: At least four
Troop size: Six to more than 100
Location: Africa

Chimpanzees are similar to gorillas. They are both apes. An ape is a primate with no tail. Chimpanzees live in the **savannas** and tropical forests of Africa. They are lively animals that live in family groups. Each troop is led by a dominant male.

Chimpanzee troops are often made up of many smaller groups. Troop members move between small groups as they please. Chimpanzees protect their home range. They often fight to defend it. They also hunt as a pack. They work together to take down their prey. Then, they share their catch with the whole troop.

Baby chimpanzees are carried by their mothers until they are six months old.

Sticking Together

Male chimpanzees stay with their birth troops their whole lives. Females leave to join other troops when they are about 11 years old.

Chimpanzee troops have two to three times as many adult females as adult males.

Bonobos

At a Glance

Troop size: 30 to more than 100
Location: Democratic Republic of the Congo (DRC)

Bonobos are found only in rain forests along the Congo River in the DRC. They are ***endangered*** *due to habitat loss and illegal hunting.*

Bonobos, another type of ape, are closely related to chimpanzees. They live in large troops that also split into smaller groups of around six to 15. Unlike gorillas and chimpanzees, however, females rule bonobo troops. They bond with their sons for life, and they never leave the troop. Females also form strong ties with each other.

All members of the troop work together for the good of the group. They look for food in small groups and share with each other. They also work together to build nests. They all sleep in the same area so they can warn each other of danger.

Bonobo troops are mainly made up of mothers and their young.

Sticking Together

Bonobos are some of the smartest animals on Earth. They share many human behaviors, such as using tools to get food. They also teach their young social skills.

Baboons

At a Glance

Number of species: Five: chacma, yellow, olive, Guinea, and hamadryas
Troop size: Five to 300
Location: Africa and Arabia

Baboons are monkeys. They have tails, unlike apes. These primates live in woodlands, hills, and savannas in dry parts of Africa and Arabia. Thick fur covers most of their bodies—except their snouts and their bottoms.

Baboons are very social animals, living in large troops. They sleep and look for food together. They also **groom** each other. Baboons work as a team to protect their troop from enemies. Unlike some other monkey species, baboons spend most of their time on the ground. They use trees to keep watch over their home range. They also eat and sleep in trees.

The average baboon troop has 50 members.

Did You Know?

The oldest baboon fossil ever found is more than two million years old.

Baboons are some of the largest monkeys in the world. They are 20 to 45 inches (50 to 115 cm) long. They weigh about 33 to 82 pounds (22 to 37 kg).

Savanna Baboons

Four of the five baboon species live south of the Sahara Desert in Africa. They are known as the savanna baboons. They live in large troops that can have hundreds of members.

Males are dominant in these troops. Each one has a **rank**. They often fight with each other to move up in rank. The highest-ranking males get to eat first. They also settle conflicts within the troop. Male savanna baboons often form friendships with females. They groom the females, bring them food, and protect them. In return, the female is more likely to mate with them.

Olive baboon

Guinea baboons

The four savanna baboon species share many of the same traits and ways of life.

Most female savanna baboons stay with their birth troops for their whole lives. Males leave their birth troops before they reach adulthood.

Hamadryas Baboons

The fifth baboon species is called the hamadryas baboon. These baboons live in the hills along the coasts of Africa and Arabia, near the Red Sea. Their large troops are made up of smaller groups called harems. Each harem has a dominant male and about four females.

Hamadryas baboons travel in smaller groups during the day. They work together to find food. Then, they gather in large groups to find water. They also gather in large groups at night. These are the times when hamadryas baboons are most **vulnerable** to predators. Gathering in larger groups gives them more protection.

A dominant male baboon might bark to warn the troop of predators, such as leopards.

*Hamadryas baboons are also called **sacred** baboons. This is because ancient Egyptians believed the animals were special to the god of learning.*

Sticking Together

Baboons can make up to 30 different sounds to communicate with each other. They also make gestures with their faces and bodies.

Baboons open their mouths wide and show their teeth when they want to look threatening.

A toothy grin shows ***submission****.*

Mandrills

At a Glance

Troop size: Usually 10 to 40, with sometimes larger groups of hundreds
Location: Western Africa

Mandrills are the largest monkey species. Males can grow up to 3 feet (90 cm) long and weigh as much as 80 pounds (36 kg).

Mandrills live in rain forests across western Africa. They spend the day looking for food on the ground. At night, they sleep in trees. These large animals live in troops made up of family groups. Each troop is led by a dominant male. Troops also come together to form much larger groups called hordes. A horde can have hundreds of members.

Male mandrills spend most of their time looking for food such as fruit, insects, and small animals. They also protect the troop. Females mainly stay in the trees with the young. Mothers feed and groom their babies. Other females help by grooming, carrying, and playing with the babies.

Dominant males often wander from the troop. However, they come back at the first sign of danger.

Did You Know?

Mandrills are known for their colorful faces. The bright colors help mandrills spot each other through the thick, dark trees. The dominant male has the most brightly colored face.

Macaques

At a Glance

Number of species: 23
Troop size: Around 20 to more than 100
Location: Asia and North Africa

Macaques can carry food, such as fruit, in their large cheek pouches.

Macaques are a type of monkey. They live on plains, in forests, on cliffs, or in rocky areas. Macaque troops can be different sizes, but most are quite large. Their home range and habits differ from one group to the next. During the day, troops break into smaller groups to search for food. They can cover more land this way.

Troops are made up of many males and females. They are led by a dominant male. Living in a troop is safer for macaques. For example, when one macaque sees a predator, they holler loudly to warn the others.

Sticking Together

Female macaques never leave their birth troops. They pass their rank on to their daughters. Lower-ranking females sometimes groom higher-ranking females. In return, high-ranking females might share their food or protect the lower-ranking females from conflicts. They might even warn each other when predators are near.

Female troop members, especially mothers and daughters, form strong bonds.

Female macaques spend a lot of time in trees, safer from predators. Males, however, spend more time on the ground.

Kangaroos

At a Glance

Number of species: Four—red, eastern grey, western grey, and antilopine
Troop size: Around 50
Location: Australia, Tasmania, and nearby islands

Kangaroos can live in a wide range of habitats. But they mainly live in woodlands, savannas, forests, and plains. Kangaroos live in large groups. Some have more than 100 members. A group of kangaroos is often called a mob, but it can also be called a herd or troop.

Did You Know?

Kangaroos have very large hind feet. They stomp their feet to warn each other when they sense danger nearby.

Members of a kangaroo troop come and go as they please. The troop is led by a dominant male. Kangaroos often form relationships within their troops. They touch noses or sniff to get to know each other. Troop members sometimes groom each other. They also help protect each other.

Kangaroos are the only large animals that hop as their main way of getting around. Their long, thick tails help them balance while they hop.

Male kangaroos box with each other to rise up in rank.

Ostriches

At a Glance

Number of species: Two—common and Somali
Troop size: 10 to 100
Location: Africa

Ostrich groups are often called flocks, but they are sometimes called troops.

Ostriches are the biggest birds in the world. They usually live in small groups of about 10 birds. But some groups have more than 100 birds. A dominant male leads each troop. There is also a dominant female, or main hen. Several other females make up the rest of the troop. Sometimes, other males join the troop to mate with the females.

Defense is one of the main reasons ostriches form troops. Their large eyes allow them to spot predators, such as lions and hyenas, from far away. The dominant male defends the troop's territory.

Sticking Together

Ostrich hens in a troop lay their eggs in the same nest. The main hen sits on the nest during the day. Then the dominant male takes her place at night. The troop works together to protect the eggs from predators. There would be fewer chicks if each hen laid her eggs in her own nest.

The main hen puts her eggs in the middle of the nest. This is the warmest, safest spot.

Ostriches are so heavy that they cannot fly. But they can run at speeds up to 43 miles per hour (70 km/h)!

Glossary

confident Having a strong belief in oneself

deter Discourage from doing something

dominant Controlling or more powerful than others

ecosystems Communities of living and nonliving things that interact with one another

endangered At risk of extinction

groom For animals, an often social behavior in which one animal cleans another's body

home range An area in which an animal lives and travels

mate One of two animals that come together to breed, or have babies

predators Animals that hunt other animals for food

prey Animals that are hunted by other animals for food

primates Mammals, such as apes, monkeys, and humans, that have shared characteristics, such as large brains and flat nails

rank A position in relation to others in a group

savannas Flat, grass-covered, and treeless areas

scat Animal droppings

species A group of living things with similar characteristics, in which two animals can have babies

subspecies Two or more distinct categories of one species

territory An area claimed and defended by an animal or group of animals

vulnerable Able to be easily hurt

Learning More

On the Web

www.kidcyber.com.au/primates
Learn all about primates and how primate species are similar and different from each other.

https://denverzoo.org/animals/mandrill
Discover all of the facts you'd ever want to know about mandrills.

https://kids.nationalgeographic.com/animals/mammals/facts/kangaroo
How much do kangaroos weigh? How do kangaroo mothers carry their babies? Learn the answers to these questions and more.

Books

Daly, Ruth. *Bringing Back the Mountain Gorilla*. Crabtree Publishing, 2020.

Lockwood, Sophie. *Baboons*. The Child's World, 2014.

Waxman, Linda Hamilton. *Ostriches: Fast Flightless Birds*. Lerner Publishing Group, 2016.

Index

About the Author

Heather C. Hudak has written hundreds of kids' books on all kinds of topics. She especially loves writing about animals. When she's not writing, she enjoys traveling and camping in the mountains with her husband and their very own animal troop, which includes five dogs and three cats!